TOP TRADE CAREERS

LAS MEJORES CARRERAS PROFESIONALES

WEB DEVELOPER
DESARROLLADOR WEB

A Crabtree Branches Book
Un libro de Las Ramas de Crabtree

Written by / Escrito por B. Keith Davidson
Translated by / Traducción de Santiago Ochoa

Crabtree Publishing
crabtreebooks.com

School-to-Home Support for Caregivers and Teachers

This high-interest book is designed to motivate striving students with engaging topics while building fluency, vocabulary, and an interest in reading. Here are a few questions and activities to help the reader build upon his or her comprehension skills.

Before Reading:

- *What do I think this book is about?*
- *What do I know about this topic?*
- *What do I want to learn about this topic?*
- *Why am I reading this book?*

During Reading:

- *I wonder why...*
- *I'm curious to know...*
- *How is this like something I already know?*
- *What have I learned so far?*

After Reading:

- *What was the author trying to teach me?*
- *What are some details?*
- *How did the photographs and captions help me understand more?*
- *Read the book again and look for the vocabulary words.*
- *What questions do I still have?*

Extension Activities:

- *What was your favorite part of the book? Write a paragraph on it.*
- *Draw a picture of your favorite thing you learned from the book.*

Apoyo escolar para cuidadores y maestros

Este libro de alto interés está diseñado para motivar a los estudiantes dedicados con temas atractivos, mientras desarrollan la fluidez, el vocabulario y el interés por la lectura. A continuación se presentan algunas preguntas y actividades para ayudar al lector a desarrollar sus habilidades de comprensión.

Antes de leer:

- *¿De qué pienso que trata este libro?*
- *¿Qué sé sobre este tema?*
- *¿Qué quiero aprender sobre este tema?*
- *¿Por qué estoy leyendo este libro?*

Durante la lectura:

- *Me pregunto por qué...*
- *Tengo curiosidad de saber...*
- *¿En qué se parece esto a algo que ya conozco?*
- *¿Qué he aprendido hasta ahora?*

Después de leer:

- *¿Qué intentaba enseñarme el autor?*
- *¿Cuáles son algunos detalles?*
- *¿Cómo me ayudaron las fotografías y los pies de foto a entender más?*
- *Vuelve a leer el libro y busca las palabras del vocabulario.*
- *¿Qué preguntas tengo aún?*

Actividades de extensión:

- *¿Cuál fue tu parte favorita del libro? Escribe un párrafo sobre ella.*
- *Haz un dibujo de lo que más te gustó del libro.*

TABLE OF CONTENTS

ÍNDICE

A community is made up of people coming together to make life better for each other.

Una comunidad está formada por personas que se unen para mejorar la vida de los demás.

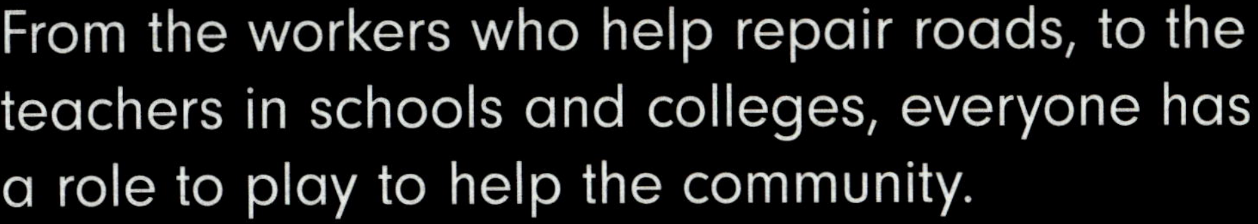

From the workers who help repair roads, to the teachers in schools and colleges, everyone has a role to play to help the community.

Desde los trabajadores que ayudan a reparar las carreteras hasta los profesores de las escuelas y universidades, todos tienen un papel que desempeñar para ayudar a la comunidad.

Web developers are important members of our **global** community.

Los desarrolladores web son miembros importantes de nuestra comunidad **global**.

WEB DEVELOPER
DESARROLLADOR WEB

Many web developers have degrees or diplomas in a computer-related field. These certifications are not requirements, but they will help start a career.

Muchos desarrolladores web tienen títulos o diplomas en un campo relacionado con la informática. Estas certificaciones no son un requisito, pero ayudan a iniciar una carrera.

Many developers go through bootcamps—weekend courses designed to teach you everything you need to know in a very short period of time.

Muchos desarrolladores asisten a *bootcamps*, que son sesiones de entrenamiento durante el fin de semana, diseñadas para enseñar todo lo que se necesita saber en un periodo de tiempo muy corto.

What's the difference between a web developer and a computer programmer? Computer programmers design software and write code. Web developers write code, but for websites only.

¿Cuál es la diferencia entre un desarrollador web y un programador informático? Los programadores informáticos diseñan software y escriben código. Los desarrolladores web escriben código, pero solo para sitios web.

SPECIAL SKILLS
HABILIDADES ESPECIALES

A basic knowledge of computers and how they operate will help, but websites have a language all their own.

Un conocimiento básico de las computadoras y de su funcionamiento será de ayuda, pero los sitios web tienen un lenguaje propio.

You will have to learn the basic computer languages of HTML (hypertext markup language), CSS (cascading style sheets), and **JavaScript**. These are the three most common languages in website development.

Tendrás que aprender los lenguajes informáticos básicos de HTML (lenguaje de marcado de hipertexto), CSS (hojas de estilo en cascada) y **JavaScript**. Estos son los tres lenguajes más comunes en el desarrollo de sitios web.

Some developers choose to work on **front-end** web development—the part that you view on your computer screen.

Algunos desarrolladores optan por trabajar en el ***front-end*** del desarrollo web, la parte que se ve en la pantalla de la computadora.

Back-end developers work on the databases and **server** side of the internet. They focus on the collection of information. **Full-stack** developers work on both sides of the website.

Los desarrolladores del ***back-end*** trabajan en las bases de datos y el lado del **servidor** de Internet. Se centran en la recopilación de información. Los desarrolladores de ***full-stack*** trabajan en ambos lados del sitio web.

Web developers work with their **clients** to design the perfect website for a business, a retail chain, a government, or a charity.

Los desarrolladores web trabajan con sus **clientes** para diseñar el sitio web perfecto de una empresa, una cadena de tiendas, un gobierno o una organización benéfica.

Each client will have different needs. The developer must be able to create a website that works for each of them.

Cada cliente tendrá necesidades diferentes. El desarrollador debe ser capaz de crear un sitio web que funcione para cada uno de ellos.

Two thirds of web usage is now done on smartphone screens. So developers have to ensure that their websites work well on mobile devices.

Dos tercios del uso de la web se hace ahora en pantallas de teléfonos inteligentes. Así que los desarrolladores tienen que asegurarse de que sus sitios web funcionan bien en los dispositivos móviles.

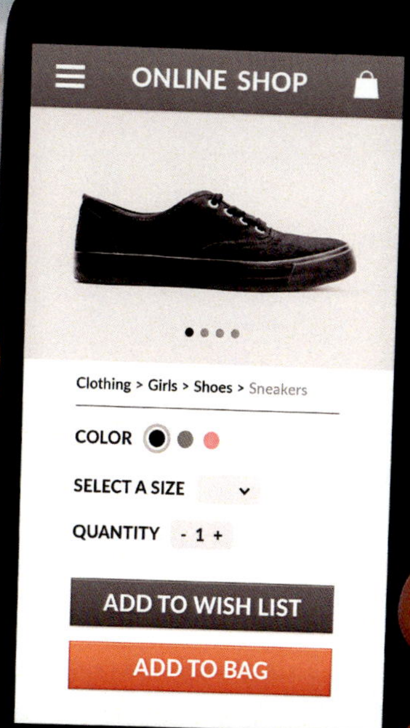

WORKING CONDITIONS
CONDICIONES DE TRABAJO

Web developers work in offices, **cubicles**, or in their own homes. If the computer they're using is a laptop, they can work almost anywhere.

Los desarrolladores web trabajan en oficinas, **cubículos** o en sus propias casas. Si la computadora que utilizan es portátil, pueden trabajar casi en cualquier sitio.

All they need for their work is a computer, Internet access, a web **browser**, and web development tools.

Todo lo que necesitan para su trabajo es una computadora, acceso a Internet, un **navegador** web y herramientas de desarrollo web.

Web developers spend a lot of time on their computers. It can lead to poor posture, muscle aches, and injuries. Work stations must be set up properly so work can be done comfortably.

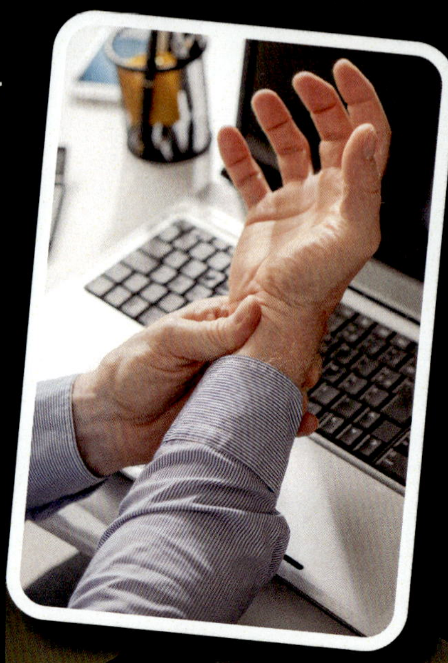

Los desarrolladores web pasan mucho tiempo en sus computadoras. Esto puede causar malas posturas, dolores musculares y lesiones. Los puestos de trabajo deben estar bien instalados para poder trabajar cómodamente.

WRONG SITTING POSTURE
POSTURA INCORRECTA AL SENTARSE

CORRECT SITTING POSTURE
POSTURA CORRECTA AL SENTARSE

Many web development companies have gyms and other activities available for their employees to keep their bodies moving during long, busy days.

Muchas empresas de desarrollo web disponen de gimnasios y otras actividades para que sus empleados mantengan su cuerpo en movimiento durante las largas y ajetreadas jornadas.

Knowing the coding languages is only half the job. Web developers work closely with their clients to make sure their design meets the customer's expectations. Developers often have to be coders, designers, AND salespeople.

Conocer los lenguajes de codificación es solo la mitad del trabajo. Los desarrolladores web trabajan estrechamente con sus clientes para asegurarse de que su diseño cumpla con las expectativas del cliente. Los desarrolladores a menudo tienen que ser codificadores, diseñadores e incluso vendedores.

The lifespan of a website is about three years. Technology is advancing so quickly that some websites may stop working properly within that time.

La vida útil de un sitio web es de unos tres años. La tecnología avanza tan rápido que algunos sitios web pueden dejar de funcionar correctamente en ese tiempo.

Web developers have to think about their clients, their clients' customers, and their competitors. There are billions of websites out there on the web. Developers have to find ways to grab attention and get views.

Los desarrolladores web tienen que pensar en sus clientes, en los clientes de sus clientes y en sus competidores. Hay miles de millones de sitios web en la red. Los desarrolladores tienen que encontrar la manera de llamar la atención y conseguir visitas.

The source code of a website can affect how it is ranked by search engines. Google ranks websites based on certain HTML elements.

El código fuente de un sitio web puede influir en la clasificación de los motores de búsqueda. Google clasifica los sitios web en función de determinados elementos HTML.

THE CHALLENGES YOU WILL FACE

DESAFÍOS A LOS QUE TE ENFRENTARÁS

Not all computer languages are created equal. They all have their issues. Developers have to know which languages are better suited to deliver the website their client needs.

No todos los lenguajes informáticos son iguales. Todos tienen sus problemas. Los desarrolladores tienen que saber qué lenguajes son más adecuados para ofrecer el sitio web que necesita su cliente.

Another challenge is the ever-changing technology. Every year there are new devices, applications, and programs. A web developer has to stay on top of all of it.

Otro reto es la tecnología en constante cambio. Cada año hay nuevos dispositivos, aplicaciones y programas. Un desarrollador web tiene que estar al tanto de todo ello.

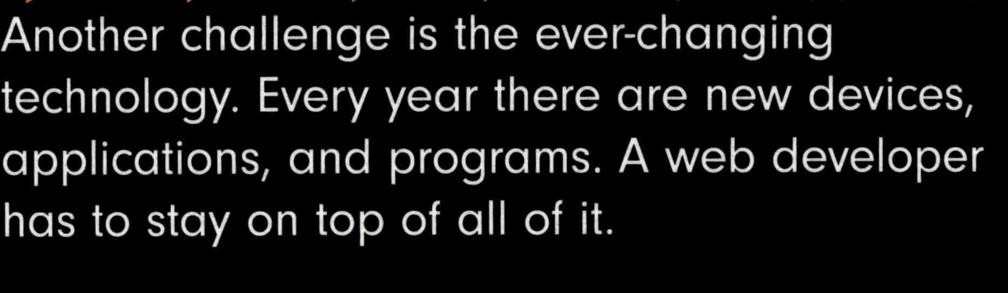

A REWARDING CAREER
UNA CARRERA GRATIFICANTE

A web developer connects people and businesses to the world.

Un desarrollador web conecta a personas y empresas con el mundo.

It is rewarding work that comes with a good salary, though the salaries vary by industry.

Es un trabajo gratificante que viene acompañado de un buen salario, aunque los sueldos varían según el sector.

Specialty	Salary Range Per Year
Publishing	$101,905 - $123,870
Computer Systems	$68,450 - $75,500
Advertising	$65,370 - $71,700
Scientific and Technical	$63,950 - $70,790

Especialidad	Rango salarial por año en dólares
Publicación	$101 905 - $123 870
Sistemas informáticos	$68 450 - $75 500
Publicidad	$65 370 - $71 700
Científico y técnico	$63 950 - $70 790

Being a web developer allows you to shape the way people see the world around them. It is a fast-paced and exciting career that is constantly growing and changing.

Ser desarrollador web te permite dar forma a la manera en que la gente ve el mundo que lo rodea. Es una carrera rápida y emocionante que crece y cambia constantemente.

It is a good career path for anyone who loves working with computers and the Internet.

Es una buena carrera para cualquiera que le guste trabajar con computadoras e Internet.

GLOSSARY

back-end (BAK-end): the part of the website that we don't see that focuses on the server and data collection

browser (BROW-sur): software that lets a user access an information system called the World Wide Web (WWW)

clients (KLY-uhnts): people who pay for the services of others

code (KOHD): program instructions for computers

cubicles (KYOO-buh-kuhlz): a small partitioned area of a room that functions as an office

front-end (FRUHNT-end): the part of the website that people view

full-stack (FUL-stak): refers to the whole website

global (GLOW-bul): relating to the whole world

JavaScript (JAH-vuh-skript): a computer code commonly used for websites

server (SUR-vur): a large computer with more storage and processing power

software (SAWFT-wair): programs designed for computers

source code (SORSS-kohd): the main code used to write a website's code

GLOSARIO

back-end: La parte del sitio web que no vemos y que se centra en el servidor y la recopilación de datos.

clientes: Personas que pagan por los servicios de otros.

código: Instrucciones de los programas informáticos.

código fuente: El código principal utilizado para escribir el código de un sitio web.

cubículos: Pequeña zona dividida de una habitación que funciona como oficina.

front-end: La parte del sitio web que la gente ve.

full-stack: Se refiere a todo el sitio web.

global: Relativo al mundo entero.

JavaScript: Código informático que se utiliza habitualmente en los sitios web.

navegador: Programa informático que permite al usuario acceder a un sistema de información llamado World Wide Web (WWW).

servidor: Una computadora grande con más capacidad de almacenamiento y procesamiento.

software: Programas diseñados para ordenadores.

INDEX

ÍNDICE ANALÍTICO

WEBSITES TO VISIT
SITIOS WEB PARA VISITAR

scratch.mit.edu

www.bls.gov/ooh/computer-and-information-technology/web-developers.htm#tab-1

https://careerfoundry.com/en/blog/web-development/what-does-it-take-to-become-a-web-developer-everything-you-need-to-know-before-getting-started

ABOUT THE AUTHOR

B. Keith Davidson has had careers in agriculture, industrial manufacturing, and the service industry. His career in education led to his current career in writing books.

SOBRE EL AUTOR

B. Keith Davidson ha hecho carrera en la agricultura, la fabricación industrial y el sector de los servicios. Su carrera en la educación lo llevó a su actual carrera de escritor de libros.

Crabtree Publishing

crabtreebooks.com   800-387-7650

Copyright © 2025 Crabtree Publishing

Written by/Escrito por: B. Keith Davidson
Designer/Diseñadora: Jennifer Dydyk
Editor/Editora: Kelli Hicks
Proofreader/Correctora de pruebas: Melissa Boyce
Translated by/Traducción de: Santiago Ochoa
Spanish-language copyediting and proofreading/
Maquetación y corrección en español: Base Tres
Coordinador de producción/Production manager:
Candice Campbell

Hardcover	978-1-0398-6933-2
Paperback	978-1-0398-6912-7
Ebook (pdf)	978-1-0398-6954-7
Epub	978-1-0398-6975-2

Printed in the U.S.A./Impreso en los Estados Unidos/CP112025

Published in Canada
Publicado en Canadá
Crabtree Publishing
616 Welland Avenue
St. Catharines, Ontario
L2M 5V6

Published in the United States
Publicado en los Estados Unidos
Crabtree Publishing
347 Fifth Avenue
Suite 1402-145
New York, NY 10016

Library and Archives Canada Cataloguing in Publication
Available at the Library and Archives Canada

Library of Congress Cataloging-in-Publication Data
Available at the Library of Congress

Photographs/Fotografías: Cover career logo icon © Trueffelpix, diamond pattern used on cover and throughout book © Aleksandr Andrushkiv, cover photo © PR Image Factory/shutterstock.com, photo at top of cover and top of title page © ronstik/shutterstock.com. Page 4 top photo © David Econopouly | Dreamstime.com, bottom photo © Firmanemmanuelle | Dreamstime.com, Page 5 © Monkey Business Images | Dreamstime.com, Page 7 © Andrey Sirant | Dreamstime.com, All other images from istock by Getty Images: Page 6 © SeventyFour, Page 8 © Photo Italia LLC, Page 9 top photo © monkeybusinessimages, bottom photo © nd3000, Page 10 © milindri, Page 11 inset photo © Minerva Studio, bottom photo © metamorworks, Page 12 © cyther5, Page 13 both photos © SeventyFour, Page 14 top photo © NicoElNino, bottom photo © kasto80, Page 15 © CarmenMurillo, Page 16 © SolisImages, Page 17 top photo © ijeab, bottom photo © Youngoldman, Page 18 top photo © thodonal, illustration © bestsale, Page 19 © howtogoto, Page 20 © NiKita Filippov, Page 21 top photo © Rostislav_Sedlacek, bottom photo © pondsaksit, Page 22 © Wasan Tita, Page 23 background photo © NanoStockk, top photo © Baks, bottom photo © mbbirdy, Page 24 © YurolaitsAlbert, Page 25 © Igor-Kardasov, Page 26 © Berezko, Page 27 © gorodenkoff, Page 28 © scyther5, Page 29 © Deagreez